'Iví' 'Axánninik
ŞAKÍSHLA
Pówki 'Axánya

A collaborative effort by the
Pechanga Chámmakilawish School, Pechanga Cultural Resources,
Julie Jackson, Neal Ibanez, Eric Elliott and Great Oak Press.

Great Oak Press,
Pechanga, California

This book satisfies several Cháammakilawish Pechanga School Language Standards.
2nd Grade:
1.9 Accurately use in daily conversations qáala 'for one inanimate thing to exist', 'áawq 'for one animate noun to exist', wónq 'for two or more inanimate things to exist', qálwun 'for two or more animate nouns to exist', and míyq 'for an uncountable amount of things to exist'.
2.4 Identify the meanings of various inflectional endings that represent tenses (-q, -an, -wun, -qu$, -lut, etc.) and be able to identify them in Luiseño texts.
2.6 Understand that any adjective describing an animate plural noun must also pluralize.
4.1 Ask and answer questions in Luiseño regarding big ideas, key details, and any visual information presented.
4.13 Read grade-level appropriate interdisciplinary Luiseño texts with proficiency and confidence in the language.
13.9 Discuss animal life cycles in Luiseño (ex. pá'kwarit 'tadpole', 'atónmay 'larva', $akíshla 'caterpillar', 'avéllaka pokí' 'pupa', pówkivish 'nymph', etc.).

For information, contact Great Oak Press, P.O. Box 2183, Temecula, California 92593,
www.greatoakpress.com.
ISBN: 978-1-942279-22-8
Photographs © / Julie Jackson
Photographs © / Adobe Stock
Translated by Neal Ibanez and Eric Elliott
Published by Great Oak Press, Pechanga, California
Printed in the United States of America

Híycha $u 'avéllaka?

'Ivíp **avéllaka**. 'Avéllakap 'eléelikat. Pówkip mómkat, chállaanamawish, pí' yúnnaantal múyyuk-kun chaparáppaat pilék.

'Óm $u 'ayálliqu$?
Háayinga wéhkun táppaat
míil **páy'wi'a'ankichum**
'avéllakam 'éxnga qálwun!

'Óm $u 'o$é''ivotaq hík páy'wi'a'ankichum 'avéllakam pomqálqalay **Californianga Kichámkunga**?

'Avéllaka potáaxaw pakáyyaat

1. Pomíilila
2. Po'é'
3. Potée'
4. Pówki pomáachangawish
5. Pówki pómkilawish
6. Po'áal
7. Poyú'
8. Popúsh potíiwila
9. Pochúungila

3

'Avéllakam 'angáayi chuxí'ma **páanil** 'axánninik. 'Avéllakam pom-míx páanilup pilék kíikat. Múyyuk poyóotu, pochuxíllax 'awá'wo wónq.

Michá' $um 'avéllakam tóvlima?

'Ivíp **$akíshla**. 'Avéllakam múyyuk-kun 'awóy tóvlima kuláawut, túu'qat popávlanga. Chíil'anik, $akíshla háqmawish, kihúutsamal pulú'chaxma. $akíshlam hamúula qwa'ún popávlay wuná' pónwun pomchíil'axvonga. 'Iví' 'axánninikup wímmaq 'avéllaka potóvlipi kuláawut mán túu'qat popávlanga $akíshla poqwá'vichupinga.

Híycha ṣu ṣakíshla?

Wám' chíil'anik **avéllaka potónmay**, potúng 'awóo ṣakíshla, 'áaxma kathúumal. Po'éekpum qwá'wuntum wóllaxma pitóo.

Hík qaléq sum $akíshlam wóllaxma?

$akíshlampum háylovi'àananga wóllaxma míykinga 9ngay teméengay 14yuk teméeyk. $akíshlampum wóllaxma pilék mómkatum pilék qaléq. $akíshlam pomtáachi qáy powáaraqala, $akíshlam **kúurinik** pomtáachiy, 'íipiti pomwólnipi míyq. 'Iví' potúng kúurish yaqáa.

Michá' $u 'axánnaxma máa'inik?

Wám' powóllax potáppaxvonga, $akíshla pemé''anik, háng'anik maríqqaxma 'avéllaka pokí', potúng 'awóo cháamamxal.

'Ivíp cháamamxal. Cháamamxalup 'owó"axma ngó'lash purúprush ʂóo'aqat 'axánninik. Cháamamxal poʂúnnga, ʂakíshlap maríqqaxma pómminik. Potáax maríqqima 'avéllaka póyk!

Chóo'on maríqqaxish
wám' potáppaxvonga,
'avéllakapo pulú'chaan.

'Angáayi, 'avéllaka pówkip
yamáqqaat, lukúppaat
potáaxawnga néshkin.

'Ehéngmayum 'axánninik, 'avéllakam pompilá'chipi míyq pomwáapaxpiy. Pilachpilá'chinik pilék, qaléqpum wáppanik pomwáapaxvotawun.

Powíilaxvotaqanik
'ayáalinik, 'avéllaka
háalan popéewi.
'Áawchorakash
yú'pan chóx'aan!

'Óm $u 'owu$ánnivotaq 'éskanaxish

'Ivá' táppaq 6

tó$ngunga 'ayóng'axnga

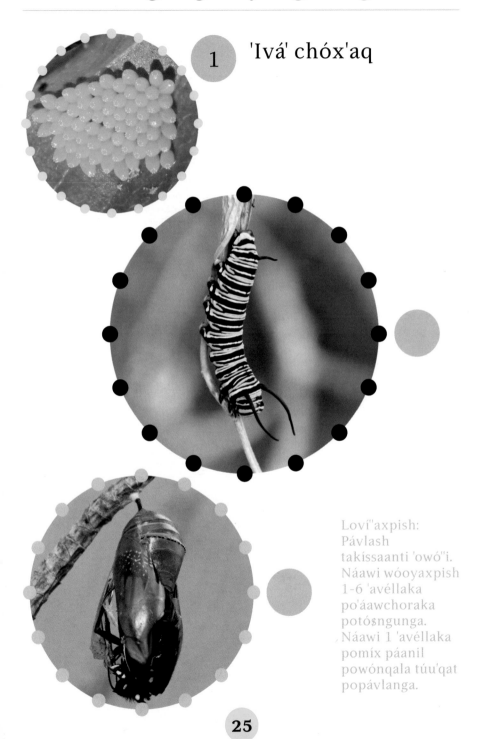

1 'Ivá' chóx'aq

Loví''axpish:
Pávlash
takíssaanti 'owó''i.
Náawi wóoyaxpish
1-6 'avéllaka
po'áawchoraka
potó$ngunga.
Náawi 1 'avéllaka
pomíx páanil
powónqala túu'qat
popávlanga.

Hík su 'óm

'Ivá' táppaq

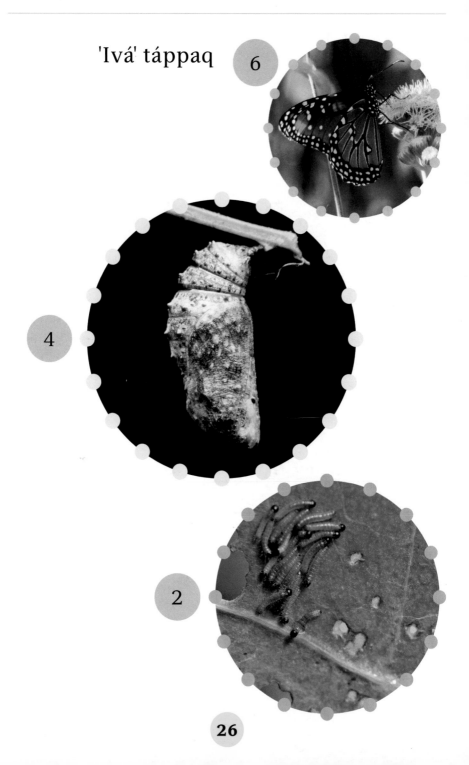

'ayáalinik húrrax?

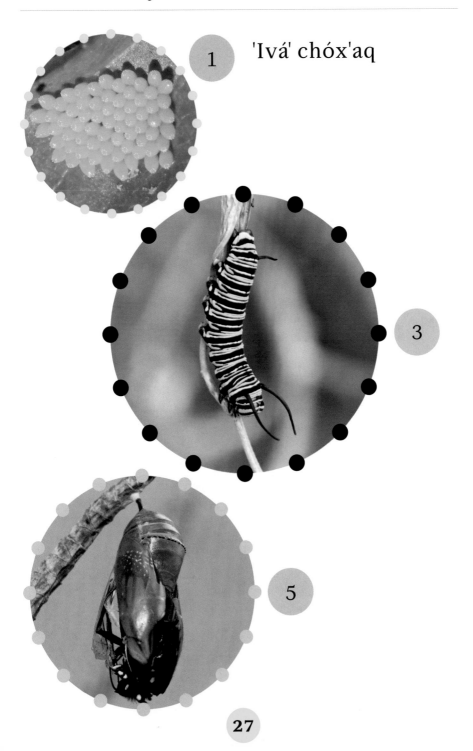

'Ivá' chóx'aq

1

3

5

English Translation

Title: How the Caterpillar Got its Wings
Subtitle: A Life Cycle Story

1. What is a butterfly? This is a butterfly. A butterfly is an insect with large, scaly, and most often, very colorful wings.

2. Did you know? There are approximately 20,000 species of butterflies in the world!

3. Can you guess how many species of butterflies live in Southern California? Answer: Estimated, 170 species.

4. No text

5. Anatomy of a Butterfly
   1. Antennae
   2. Legs
   3. Abdomen
   4. Hindwing
   5. Forewing
   6. Thorax
   7. Head
   8. Eyes
   9. Proboscis

6. No text

7. Butterflies start life as an egg. Butterfly eggs are very small and come in many different shapes and colors.

8. Where do butterflies lay their eggs? Butterflies usually lay their eggs on the leaves of plants and trees. When the eggs hatch, out pops a tiny and hungry caterpillar. Caterpillars will start by eating the leaf it was born on. This is why it is important for the butterfly to lay her eggs on a leaf that the caterpillar will eat.

9. No text
10. No text
11. What is a caterpillar? This is a caterpillar. A butterfly larva or caterpillar looks like a small worm when it first hatches. In this stage all they do is eat and grow.
12. How fast do caterpillars grow? It takes 9 to 14 days for a caterpillar to be fully grown. Caterpillars grow very big, very fast. Since caterpillars don't have skin that stretches as they grow, they must shed their skin and grow new skin. This is called molting.
13. No text
14. What happens next? When the caterpillar has finished growing, it hangs upside down and forms itself into a pupa, also know as a chrysalis.
15. No text
16. No text
17. This is a chrysalis. A chrysalis acts as a hard protective case. Inside the chrysalis, the caterpillar is undergoing an extraordinary transformation. A butterfly is forming!
18. When all of the changes inside the chrysalis are complete, a butterfly will emerge.
19. At first, the butterflies' wings are soft and folded against its body.
20. Like birds, butterflies have to learn how to fly. It takes a lot of practice, but they learn quickly.
21. No text
22. No text
23. Once the butterfly has mastered flight, it will search for a mate and the life cycle begins again!
24-25. Can you put the pictures in the correct order?
       Ends Here
       Starts Here

Activity: Use tracing paper. Fill in the numbers 1-6 in the order of the butterfly life cycle—1 beginning with the butterfly eggs on the leaf.

26-27. How many did you get correct?

Ends Here

Starts Here